The Future of Artificial Intelligence

Impacts and Opportunities

Table of Contents

1. Introduction 1

2. Introduction to Artificial Intelligence: A New Frontier 2

    2.1. The Birth of AI 2

    2.2. Early Obstacles and AI Winter 3

    2.3. Laying Foundations 3

    2.4. Arrival of Neural Networks 3

    2.5. Modern AI: Rise of Machine and Deep Learning 4

3. Understanding AI: From Basic Concepts to Advanced
Technologies 5

    3.1. The Basics of AI 5

    3.2. AI Technologies: Machine Learning and Deep Learning 6

    3.3. The Power of AI 7

    3.4. Ethical Concerns of AI 7

4. The AI Revolution: Current Applications and Real-World
Scenarios 9

    4.1. Machine Learning and Analytics 9

    4.2. AI in Healthcare 10

    4.3. AI in Autonomous Vehicles 10

    4.4. AI in Retail and E-commerce 11

    4.5. AI in Education 11

5. Future Projections: Where is AI Heading? 13

    5.1. The Pervasive Future of AI 13

    5.2. Transforming Job Markets and the Economy 14

    5.3. Healthcare: A Life-Saving Revolution 14

    5.4. Education: Personalized Learning and Teaching Aids 15

    5.5. Ethical Challenges Ahead 15

    5.6. The Road Ahead: AI Policy and Regulation 16

6. Impacts on Society: Ethics, Privacy, and Security 17

6.1. Ethical Considerations in AI 17

6.2. Privacy Concerns in the Age of AI 18

6.3. AI and Security 19

7. AI and the Economy: Disruption and Transformation 20

7.1. Disruptive Innovations: AI, Economic Growth, and
Productivity 20

7.2. The Changing Nature of Work: Job Creation and
Transformation 21

7.3. Regulatory Challenges and Fairness Considerations 22

7.4. Inequality Concerns: Mitigation Strategies 22

7.5. Conclusion 23

8. Business Opportunities: Harnessing AI for Competitive
Advantage 24

8.1. Integrating AI and Strategic Planning 24

8.2. AI in Supply Chain and Operations 25

8.3. Process Automation and Cost Optimization 25

8.4. AI-Driven Marketing and Personalization 26

8.5. Enhancing Customer Service with AI 26

9. Career Prospects in the AI Era 28

9.1. Jobs Directly Related to AI 28

9.2. The Importance of Soft Skills in the AI Era 29

9.3. Educational Opportunities in AI 29

9.4. AI Impact across Various Industries 30

10. AI and the Individual: Everyday Implications and Interactions . 32

10.1. Understanding AI in Everyday Life 32

10.2. AI and Personal Health Management 33

10.3. AI in Education 33

10.4. AI and Personal Security 34

10.5. The Possibilities and Challenges Ahead 34

11. Conclusion: Preparing for the AI-Driven World 36

11.1. Embrace the Changing Reality 36

11.2. Fostering Digital Literacy and AI Education 37

11.3. Nurturing AI Innovation 37

11.4. Ethics and Regulations in an AI World 38

11.5. Mitigation of AI Induced Displacement 38

11.6. Collaborative AI Advancements 38

Chapter 1. Introduction

Welcome to this Special Report on the fascinating topic: "The Future of Artificial Intelligence: Impacts and Opportunities." It's no secret that the world of AI is complex and rapidly changing, which may make the topic seem daunting; however, we've meticulously crafted this report to distill those intimidating complexities into clear, digestible insights. In this comprehensive review, we traverse the landscape of AI's future, exploring its potential impacts on society, industry, and personal lives, as well as the plethora of prospects it provides. Neither overly technical in the presentation nor simplified beyond usefulness, it's a complete, balanced exploration of the AI universe that anyone can understand and enjoy. Prepare yourself for a journey into the future, one that informs, educates, and sparks imagination, resonating the possibilities this technology holds. We want to equip you with the knowledge and context to understand, navigate, and perhaps even shape the rapidly evolving world of AI. Get ready to dive into this incredible adventure filled with striking revelations and exciting opportunities!

Chapter 2. Introduction to Artificial Intelligence: A New Frontier

Artificial Intelligence (AI) is not a new concept. Its roots drive deep into our cultural, scientific, and philosophical histories. Nonetheless, the past decades have witnessed monumental leaps in its capabilities, reaching a scale of influence that its early proponents could only dream of. Today, AI has become an integral part of our everyday lives and shows no sign of slowing down its rapid, transformative growth.

AI's present prowess and future potential is largely owing to immense advances in its underlying technologies and methodologies. At its core lies machine learning (ML). By leveraging algorithms and statistical models, ML allows computer systems to accomplish tasks and make predictions or decisions without explicit programming. This differentiates ML (and by extension, AI) from traditional software, which relies on hand-coded rules and heuristics.

2.1. The Birth of AI

AI's journey began in earnest in the mid-20th century, fueled by groundbreaking work from scholars like Turing and others. In the 1956 Dartmouth Workshop, organized by Marvin Minsky and John McCarthy, the term 'Artificial Intelligence' was coined. This seminal event marked the official beginning of AI as a field of study. Early optimism led to the expectation of machines outperforming humans in every cognitive task within a generation. However, these high hopes were faced with reality as researchers discovered the complexity of bridging the gap between human and machine intelligence.

2.2. Early Obstacles and AI Winter

The technical and conceptual challenges in the nascent field led to reduced enthusiasm and funding, ushering in periods known as "AI winters." Several influential AI projects faced difficulties due to the size, diversity, and complexity of human knowledge—hard to replicate within the constraints of computational resources and methods then available. These included problems in natural language understanding, computer vision, and common-sense reasoning. Additionally, the emergence of Noam Chomsky's generative grammar refuted initial hopes of using simple statistical or rule-driven methods to parse human languages.

2.3. Laying Foundations

Despite these setbacks, the early phases were not fruitless. Essential landmarks were achieved, such as developing logical formalisms (like predicate calculus) for common-sense knowledge. Key elements of problem-solving strategies were also developed in these early years, such as means-ends analysis and general problem-solving architectures like the General Problem Solver (GPS). Early rule-driven AI programming languages, like IPL and LISP, were also introduced, which laid the groundwork for future developments.

2.4. Arrival of Neural Networks

The first models of artificial neurons date back to as early as the 1940s, with the work of McCulloch and Pitts. Rosenblatt's perceptron model in the late 1950s further pioneered this area. However, it faced criticism and went largely dormant until the mid-1980s, when researchers developed the backpropagation technique for training multilayer neural networks. This ushered in a new era for artificial neural networks, which, coupled with the emergence of digital computers, led to much of the success we see in the present day AI, specifically in deep learning.

2.5. Modern AI: Rise of Machine and Deep Learning

The trend in AI progressed from Hand-Crafted Knowledge era of 1956-1986, to the Machine Learning era of 1986-2012, and then to the present Deep Learning era, commencing from 2012. The ever-increasing volume of data paired with advancements in computational power has steered AI towards ML-based methods. The staggering capabilities of deep learning, an ML subset involving artificial neural networks with multiple hidden layers, have sparked a massive wave of innovation in the AI sphere.

Deep learning networks learn to recognize patterns in raw data, addressing many of the limitations of earlier techniques by enabling machines to process large datasets and perform complex tasks kin to human capabilities. These include, yet not limited to, identifying images, recognizing speech, processing natural language, and even mastering sophisticated board games.

Now, as you read this report, AI stands at a crucial juncture. While there is immense optimism for this technology's potential and impact, there are also crucial challenges and questions concerning ethics, regulations, job displacement, and inequality. As we delve deeper into this report, it's essential to understand that AI is not just binary code running on silicon—it also embodies a broader socio-technical system. The dawning era of AI will fundamentally reshape our society, culture, and individual lives, forever transforming humanity's journey into the future.

Chapter 3. Understanding AI: From Basic Concepts to Advanced Technologies

In this journey of understanding the intricate realm of AI, we start with the most fundamental question - What is AI? Ultimately, Artificial Intelligence (AI) is about designing machines that can think and learn like humans. This broad and ever-evolving field is the confluence of computer science, mathematics, psychology, neuroscience, cognitive science, and many more. The terms AI and machine learning (ML) are often used interchangeably, although they are not the same. Machine learning is a subset of AI, and it consists of the techniques that enable computers to figure things out from the data and deliver AI applications.

3.1. The Basics of AI

Why does AI matter? It is no exaggeration to say we are living in times where AI technology underpins much of our daily lives, even if we aren't always aware of it. From personalized recommendations when online shopping to voice assistants like Alexa and Siri, AI is increasingly part of our daily routines. It's revolutionizing industries, improving efficiency, and even leading us to the brink of self-driving cars.

AI typically requires a large amount of data to learn from and deliver accurate predictions or decisions. For example, an AI system learning to identify cats in images would need thousands, or even millions of images of cats to learn from. This process of learning from vast datasets is known as 'training' the AI, and this forms the basis of many machine learning models. Once trained, these models can be deployed to perform various tasks.

Let's consider the AI behind Google's search algorithms. Well-tuned AI systems scan the entire internet to present you with the best results for your query and even predict what you might type next. They study your patterns, refine their algorithms, and serve you custom-tailored search results, all in the blink of an eye.

3.2. AI Technologies: Machine Learning and Deep Learning

Machine Learning (ML) is the most common way of achieving AI. By training an algorithm on a large dataset, it can make predictions or decisions with little human intervention. In its simplest form, machine learning can be categorized into three types:

1. Supervised learning: The machine learns a function that maps an input to an output based on example input-output pairs.

2. Unsupervised learning: The model is given a set of data and asked to find patterns within it.

3. Reinforcement learning: Algorithms learn to perform a task over many steps, improve over time, and aim to maximize some type of reward.

A famous example of reinforcement learning is Google's DeepMind creating an AI that learned to play the game Go and defeated the world champion in 2017.

Deep Learning (DL) is a more specific method of machine learning, based on artificial neural networks. Imagine the human mind, with billions of neurons firing and interconnecting. AI thinkers wondered if they could recreate this in machine form – that's exactly what deep learning attempts to do.

Deep learning architectures like Convolutional Neural Networks (CNNs) specialize in image recognition tasks, while Recurrent Neural

Networks (RNNs) excel in sequential data such as language.

3.3. The Power of AI

Perhaps the most fascinating aspect of AI is how it can, and is, beating humans at our own games – literally and metaphorically. From diagnosing diseases with better accuracy than doctors to defeating human world champions at chess, Go, and even televised quiz shows, AI is proving its incredible potential.

Modern AIs have achieved human-level performance in a wide range of games, as well as tasks such as translating languages and even writing text. This report, for example, could have been written by a sufficiently advanced AI.

But the power of AI isn't just about beating high scores and winning games, it's about improving the efficiency, accuracy, and scalability of our solutions to real-world problems.

Healthcare has seen a slew of AI-powered innovations, from AI algorithms capable of early disease detection to sophisticated robots assisting surgeons in complex procedures. In finance, AI is used for automating trading, predicting market fluctuations, analyzing and predicting customer behaviors, and advancing cybersecurity.

3.4. Ethical Concerns of AI

With great power comes great responsibility. While AI brings about many advantages, its impact on our society, economics, and individual lives necessitates crucial ethical questions. These range from ensuring AI doesn't perpetuate human bias in its decision-making to ensuring all members of society can benefit from AI advancements.

It's important that policymakers, researchers, and the general public are involved in conversations about the ethical implications of AI, data privacy, job automation, and many other areas. These conversations are critical moving forward as we aim to harness the power of AI responsibly.

As we finish this section, it's key to remember that AI will continue to evolve. Thus, understanding its basics and advanced technologies is a dynamic process that involves continuous learning and adaptation.

The road ahead is exciting, challenging, and at times uncertain. However, armed with understanding and foresight, we can navigate the twists and turns, and brace ourselves for an AI-driven future. This is why understanding AI is not just about knowing its technicalities but also about appreciating its potential, reckoning with its challenges, and acknowledging the responsibilities it entails.

Throughout this report, keep in mind that AI is not a single technology, but a collection of technologies that are applied in a wide range of contexts. It's not something monolithic or static – that's part of what makes it so exciting and so crucial to understand.

Chapter 4. The AI Revolution: Current Applications and Real-World Scenarios

Artificial Intelligence (AI) is not just a technology reserved for futuristic fantasies or confined to a laboratory; it's becoming an integral part of our everyday lives. From smartphones to smart homes, from healthcare to retail, the applications of AI are vast and growing exponentially. In this chapter, we delve into some of the prominent real-world settings where AI is already making a significant impact and explore how AI advancements could revolutionize these spaces in the coming years.

4.1. Machine Learning and Analytics

Machine Learning—a core branch of AI—provides systems with the capability to learn autonomously and improve from experience. This technology is indispensable today, backing the vast perceptiveness of modern big data analytics. For instance, Machine Learning powers the recommendation algorithms found in many digital platforms. Whether it's the suggested videos on your YouTube feed, personalized shopping recommendations on Amazon, or the pages Facebook promotes, all these analyses are fueled by Machine Learning algorithms studying vast user data to predict preferences accurately.

In the business world, Machine Learning drives insightful decision-making, risk minimization, and operational optimization through predictive analytics. Fed with past data, these algorithms can model various scenarios, anticipating future outcomes. Consequently, they have become valuable instruments for sectors like finance where they are utilized for fraud detection, credit scoring, investment modeling, and more.

As Machine Learning evolves, we anticipate greater leaps in data-driven decision-making, enhancing efficiency and effecting comprehensive personalization in various sectors.

4.2. AI in Healthcare

AI has made significant inroads into healthcare, radically transforming diagnosis, treatment, research, and patient care. Machine learning models are trained to diagnose diseases, sometimes with greater accuracy than human practitioners. For instance, AI systems can analyze medical imaging to detect signs of diseases like cancer, often in their early stages, leading to quicker and more effective treatment.

AI technologies also aid in drug discovery and development, a typically time-consuming and expensive process. Leveraging big data analytics, AI can rapidly sift through vast arrays of chemical compounds, genetic information, and clinical trials data to predict a drug's efficacy and safety profile.

Looking into the future, we envision AI-powered personalized medicine, accurately mapping an individual's health profile to tailor therapeutic solutions. Additionally, AI would facilitate remote patient monitoring, predictive health care, and automated nursing assistants, refining the quality of care, and increasing its accessibility.

4.3. AI in Autonomous Vehicles

Driverless cars have stopped being just part of science fiction. AI plays a critical role in this burgeoning field, powering the multitude of sensors that detect and interpret the car's surroundings, making accurate driving decisions. Assisted by machine learning algorithms, these AI systems improve over time, advancing the autonomous vehicles' safety and efficiency.

The impacts of fully autonomous cars could be far-reaching, from reductions in traffic congestion and pollution to potential transformations in urban planning and real estate. In a fully driverless future, daily commutes could turn into productive time or leisure, drastically altering our relationship with travel.

4.4. AI in Retail and E-commerce

AI has drastically recast the retail and E-commerce landscape, providing customer-centric shopping experiences. Machine learning algorithms offer personalized recommendations based on individual buying habits and preferences, enhancing customer satisfaction and boosting sales.

AI also improves supply-chain management, optimizing inventory levels, and ensuring timely delivery through predictive analysis. Chatbots and AI-assisted customer service improve interaction and response times, enhancing customer engagement.

Envisioning the future, brick-and-mortar stores could turn into smart stores with real-time customization, like changing prices on demand, or interactive fitting rooms with virtual assistants. AI could also reshape the overall shopping journey with technologies like voice assistants and visual search, making shopping more intuitive and enjoyable.

4.5. AI in Education

AI in education is an emerging, yet rapidly growing, field. AI-powered education platforms can adapt to an individual's learning pace and style, providing personalized content and tasks, enhancing learning outcomes.

Automatic grading systems not only reduce teachers' workloads but also provide quicker feedback, enabling students to learn from their mistakes promptly. Furthermore, predictive analytics can forecast students' future performance, helping educators spot and address issues early.

Looking ahead, AI could transform remote learning, making it more interactive and engaging. Potentially, it could bring quality education to remote and underserved areas, democratizing access to knowledge.

In conclusion, these are merely a few arenas where the AI revolution is taking place. AI's ultimate impact is bound to be greater and, quite possibly, in ways that we are yet to envision. As AI continues to grow more sophisticated, it will, no doubt, continue to enhance and redefine our lives in extraordinary ways. Just as the Internet revolutionized the late 20th century, the AI revolution seems set to reshape the 21st. We stand right at the cusp of this exciting era, and how we adapt and harness this wondrous technology could well define our collective future.

Chapter 5. Future Projections: Where is AI Heading?

Artificial Intelligence (AI) can seem like something from the realm of science fiction until we remember that we're already interacting with it daily. From smartphone virtual assistants to targeted online advertisements, AI is right here in our present, even as it continues to drive ever closer towards a rapidly approaching future. This chapter explores projections for this future, attempting to answer the critical question - where is AI heading?

5.1. The Pervasive Future of AI

In the future, our interactions with AI could become even more pervasive, infiltrating each aspect of our lives. We might see AI deeply integrated within homes, offices, transport systems, and public services. This could lead to more streamlined operations and efficient services. For example, AI might manage food storage and shopping lists within our homes, organise our work tasks for optimal productivity or regulate traffic to minimise congestion on our roads.

However, this widespread deployment of AI also calls attention to concerns such as privacy and data security. As AI systems continue to learn and evolve further, they would need more accumulated data, which could raise questions about how personal information is stored, shared, and utilised.

5.2. Transforming Job Markets and the Economy

Another important area where AI is heading is the transformation of job markets and economic structures. AI machines and automation are becoming increasingly capable in performing tasks that were once unique to humans, such as decision-making, problem-solving, and creative thinking.

For certain job roles, this may lead to the displacement of human workers. However, while this can seem threatening, it doesn't necessarily mean an end to human jobs. Instead, we may see a shift in the type of work humans do. There will be increasing demand for roles that involve the supervision, construction, and maintenance of AI systems. Furthermore, with AI shouldering repetitive tasks, humans could be free to focus on complex, creative, and more intellectually stimulating work.

Moreover, AI could spur significant economic growth. As industries embrace automation, they may experience cost-efficiency and productivity gains, leading to wealth creation and possible economic expansion.

5.3. Healthcare: A Life-Saving Revolution

AI could revolutionize the healthcare industry by providing high precision diagnostics, personalized treatment, and more proactive health management. AI algorithms are currently being designed and tested to detect diseases such as cancer and diabetes at an early stage, thereby dramatically improving the chances of successful treatment.

In the future, AI could not only diagnose but also offer treatment suggestions based on a patient's individual genetic make-up, lifestyle, and medical history. Moreover, AI deployed in wearables could monitor vital signs, predict potential health risks, and provide real-time advice for a healthier lifestyle. On a larger scale, AI could enable population health management and epidemic control by analysing patterns and predicting disease outbreaks.

5.4. Education: Personalized Learning and Teaching Aids

AI can effectively personalize learning experiences based on a student's unique learning style, pace, and interests. This could lead to far-reaching transformations in education, with AI tutors providing customized lesson plans, and teachers leveraging AI tools to identify student strengths and weaknesses and provide targeted support.

In addition, AI can also offer solutions for the democratization of education. AI-based applications and online learning platforms could make quality education accessible to remote or underserved communities.

5.5. Ethical Challenges Ahead

While AI holds impressive potential, it also brings about ethical instabilities. For instance, there are unanswered questions regarding the responsibility and accountability for AI decisions, which are often called the "black box" dilemma. If an AI makes a consequential decision that negatively impacts individuals or society, then who is held accountable?

Another key concern is bias in AI decision-making. As AI learns from historical data, they may inadvertently learn and perpetuate societal biases—for instance, in areas such as job recruitment, credit approval, and law enforcement. The extent of AI's potential societal impact therefore necessitates the creation of an ethical and legislative framework surrounding its use.

5.6. The Road Ahead: AI Policy and Regulation

Given AI's pervasive penetration into everyday life, it seems inevitable that a regulatory framework around AI will need to evolve. Governments across the globe are outlining strategic policies for AI's development, use, and education. Regulation could play a key role in ensuring that AI develops in a way that is beneficial to society, maintains human dignity, and respects privacy and data ownership.

International collaboration will be essential on this front, as data and AI technologies traverse national boundaries. Global categories of AI applications could be established, with varying degrees of regulation based on potential risk. This can ensure safety and accountability without stifling innovation.

To conclude, the future of AI is full of opportunities for making our societies more efficient, productive, and accommodating. However, it is also fraught with ethical, legal, and societal challenges. It's a future that calls for informed decisions, proactive policy-making, and responsible AI development and deployment. The road ahead is indeed challenging, but also exciting, and navigating it successfully may well define the course of human progress in the years to come.

Chapter 6. Impacts on Society: Ethics, Privacy, and Security

With AI technologies becoming increasingly influential in daily life, it is imperative that we examine their societal implications, addressing vital issues such as ethics, privacy, and security.

6.1. Ethical Considerations in AI

Ethical questions surrounding AI are as diverse as the technology itself. From fundamental questions about AI's impact on jobs to the more subtle implications of algorithmic bias, the ethical fabric of AI is a complex, multifaceted canvas.

Machine learning, a subset of AI, presents distinct ethical challenges. Algorithms learn from data—thus, if biased or flawed data is provided, it significantly impacts the algorithm's performance. For example, facial recognition software has shown less accuracy in identifying people from minority ethnicities, raising concerns about racial bias.

Bias in AI is not only prevalent in facial recognition systems but also found in areas like credit scoring, where decisions could be influenced by factors not accurately correlating with creditworthiness—creating a risk of financial exclusion for disadvantaged groups.

Moreover, the obscure complexity of many AI systems poses ethical questions. If an AI system causes harm, who is to blame? The manufacturer? The programmer? Or the AI itself?

Additionally, Artificial General Intelligence (AGI), machines with cognitive capabilities equal to or beyond human intelligence, introduces new ethical quandaries. Would such AI qualify for rights, or be allowed to make critical decisions without human intervention?

6.2. Privacy Concerns in the Age of AI

As AI technologies become more sophisticated, they pose new challenges to maintaining personal privacy. Many AI systems collect, analyze, and process vast amounts of personal data to make accurate predictions and support decision-making. This can range from "big data" analytics used in marketing to facial recognition technologies in surveillance.

For example, AI algorithms curate the news, advertisements, and movie recommendations we see online, often leveraging extremely detailed personal data. While these services can be convenient, they're also creating a substantial digital footprint of each user, raising privacy concerns. Should companies and governments have the right to access, retain, and use this sensitive data?

Further complicating this landscape is the advent of AI systems capable of producing realistic synthetic media, commonly termed "deepfakes." They can manipulate digital content at an unprecedented level, fabricating convincing videos, voices, and photos. The potential misuse of this technology for violating privacy and spreading misinformation is an increasingly important issue.

In response, privacy-enhancing technologies (PETs), including differential privacy and federated learning, are being developed to ensure the benefits of AI can be gleaned without sacrificing privacy.

6.3. AI and Security

The rapid development and deployment of AI also pose significant security challenges. Cybersecurity has become a major global concern, and AI can be used both as a tool for defense and a weapon for attack.

On one hand, AI can significantly improve cybersecurity by rapidly detecting fraudulent activities or anomalies, predicting potential threats, and automating responses. On the other, the same technology can be exploited by malicious actors to conduct cyber attacks more efficiently. For example, AI can automate the creation of phishing emails or assist in more sophisticated attacks such as those involving deepfakes.

Physical security is also an area to consider. Autonomous weapons, vehicles, or drones—controlled by AI systems—could be hacked or malfunction, causing harm.

In conclusion, AI has profound ethical, privacy, and security implications for society. While its benefits can be tremendous, the risks must be carefully managed. Both public discourse and legislative measures should aim to guide the development and use of AI aligning it with societal norms and values. This theme carries us forward into our next expedition on the industrial impacts of AI.

Chapter 7. AI and the Economy: Disruption and Transformation

AI's influence on the economy is as profound as it is pervasive. Leveraged correctly, it offers transformative potential that extend far beyond conventional norms, engendering much-needed upheaval in landscapes where stagnation has often been the status quo.

7.1. Disruptive Innovations: AI, Economic Growth, and Productivity

AI is often likened to general-purpose technologies such as electricity or the combustion engine - fundamental innovations that sparked revolutions and led to tremendous productivity gains across multiple sectors. An AI-driven economy promises similar growth potential.

Current research suggests that AI could substantially increase economic growth and efficiency. According to a recent PwC study, AI implementations across world economies could contribute up to $15.7 trillion to the global GDP by 2030, driven primarily by boosts in productivity and consumption-enhancing effects.

Productivity gains arise from automating processes (thereby saving labor) and augmenting workers' capabilities. Automated systems can perform tasks faster and more precisely than humans, whereas augmentation enhances human decision-making, thus improving outcomes.

For instance, consider how AI automation has transformed the manufacturing sector. Traditional assembly lines, once completely manned, are now complimented by smart factories where AI-controlled machines and robots ensure round-the-clock productivity with unparalleled precision. AI goes beyond automation, extending new analytic capabilities to decision makers, facilities managers, and logistics coordinators, providing insights that were once unattainable.

The consumption side improvements are largely associated with the creation of new AI-enhanced products and services. For example, AI-assisted healthcare has led to applications that optimize patient treatment by providing more accurate diagnoses, potentially increasing life expectancy and quality of life. This creates new value, and therefore increases consumption.

7.2. The Changing Nature of Work: Job Creation and Transformation

While AI is often criticized for automation-related job losses, it is equally important to acknowledge the job creation potential presented by AI. Various sectors such as Technology, Healthcare and Education, identifying and analyzing data patterns, creating and maintaining AI systems - these are all jobs that didn't exist before the advent of AI. These newly formed roles don't merely replace the old; they also tend to be higher-skilled and better-paying.

AI also propels job transformation, which is distinct from job creation. Picciano (2019) argues that "rather than AI replacing most jobs, a more realistic scenario is that AI will instead change the tasks within jobs". Humans will work with machines, where AI systems automate routine tasks, freeing up time for workers to focus on complex problem-solving and creative tasks. This hybrid model, termed "collaborative intelligence," is likely to redefine multiple jobs, from healthcare and teaching to transportation and construction.

7.3. Regulatory Challenges and Fairness Considerations

AI-algorithms can also disrupt critical societal facets, including regulatory systems and fairness frameworks. For instance, in finance, where investment decisions, risk profiles and more are being digitally managed by AI, complexities of algorithmic bias, transparency, and cybersecurity warrant stringent regulation.

Similarly, questions of algorithmic inequality arise when AI systems inadvertently reinforce human biases. For example, a hiring algorithm trained on a data set featuring gender bias might replicate this bias in its screening process - a clear violation of equal opportunity statutes that our society upholds.

Therefore, it's crucial for governments and organizations alike to proactively foresee and address these challenges. The future holds for plenty of regulated self-learning algorithms and AI audits to maintain a fair, transparent digital economy.

7.4. Inequality Concerns: Mitigation Strategies

AI may exacerbate income inequality if the benefits accrue largely to those who own AI-powered firms or technologies. To mitigate the risk, public policy must facilitate fair distribution of AI's economic benefits. For instance, policies could encourage AI-driven firms to distribute their profits more broadly, perhaps through stakeholder models or worker dividends.

Further, policies should also focus on education and re-skilling to ensure that the workforce adapts to the changes brought by AI, especially since low and medium-skilled workers may be the most vulnerable to job displacement caused by AI.

7.5. Conclusion

AI presents a bouquet of possibilities for the economy, heightened productivity, efficiency, and the birth of innovative services among them. While these advantages are enormous, they come bundled with an array of challenges. It's up to policy makers, businesses, and society as a whole to ensure that we navigate this AI-driven economic transformation in a way that the benefits are maximized, and equitably distributed, and the risks are minimized. The future of artificial intelligence in the economy will be as we shape it; let us shape it thoughtfully—for the betterment of all.

Chapter 8. Business Opportunities: Harnessing AI for Competitive Advantage

Typically, you'd walk into a retail store, and the onus would be on you to find the product you need, occasionally aided by the store's assistants. However, the future of this scene could be dramatically revolutionized by artificial intelligence. Imagine stepping into a store where the smart systems recognize you via facial recognition, understand your past purchases and preferences through data analysis, and immediately offer personalized product recommendations and optimal paths to reach them.

In this way, AI fundamentally changes the formulae as we know them, presenting unique opportunities for businesses to gain a competitive edge. In the following sections of this chapter, we will elaborate on how AI can be harnessed for a unique advantage in various domains of business.

8.1. Integrating AI and Strategic Planning

The ability of AI to analyze substantial volumes of data in split seconds will lend an enormous helping hand to strategic planning. It can sift through and synthesize company data – financial reports, market analysis, client information - reducing the turnaround times for building strategic models significantly.

Moreover, predictive analytics supported by AI can help in forecasting market trends, customer behavior, and potential business risks. By leveraging these insights, companies can build innovative, robust, proactive strategies that keep them one step ahead in the competitive landscape.

8.2. AI in Supply Chain and Operations

Operational efficiency is a key determinant of a business's success, and AI can significantly optimize this aspect. It can monitor real-time data, enabling companies to anticipate logistical hiccups and rectify them promptly. Moreover, the technology can automate and optimize inventory management, reducing human error, augmenting predictability, and saving significant costs.

Predictive maintenance, powered by AI, could reduce unexpected operational disruptions. By analyzing patterns among numerous parameters like machine vibrations, temperature, and others, AI can predict potential breakdowns and trigger maintenance activities even before any issue becomes apparent, reducing downtime and costs related to breakdowns.

8.3. Process Automation and Cost Optimization

Automation, powered by AI, has already left its mark in various business functions like customer service, HR, and finance. AI-powered chatbots and virtual assistants can now handle many tasks like general customer inquiries, scheduling meetings, invoice processing, payroll, and much more, liberating employees to focus on more complex tasks.

This automation could lead to large-scale productivity enhancement and significant reductions in operational costs. Importantly, it can also bring down the instances of human error and inconsistencies, ensuring a higher quality of output.

8.4. AI-Driven Marketing and Personalization

Consumers today expect personalized experiences, making it a critical aspect of contemporary marketing strategies. AI excels at automating and maximizing personalization. By analyzing massive troves of customer data - including past interactions, product preferences, demographics - AI can create hyper-personalized communication for each customer, enhancing customer satisfaction and fostering brand loyalty.

Moreover, AI can automate the scheduling and distribution of marketing content based on user behavior and responses to previous communications. This could ensure that the right message reaches the right person at the right time, increasing the effectiveness of marketing campaigns.

AI can also provide valuable insights into the effectiveness of marketing initiatives, highlighting gaps and areas of success, thereby refining marketing methods, and ensuring resources are used optimally.

8.5. Enhancing Customer Service with AI

Artificial intelligence-powered chatbots and virtual assistants can handle customer inquiries 24x7, providing instant replies and constant connectivity. This would ensure a quicker resolution of customer issues, foster better relationships, and improve customer retention.

AI can also help in predicting and understanding customer behavior. It allows businesses to tailor interactions and resolve potential complaints even before they arise, empowering businesses to offer superior customer service.

While this chapter provides a general overview of how businesses can leverage AI to optimize operations, drive growth, and secure a competitive edge, the opportunities offered by AI are not limited to these areas. It is a continually evolving field with significant potential to reshape business practices, making it crucial for modern businesses to understand and appreciate the technology's potential depth and breadth. Innovation in AI will undoubtedly continue to present exciting new opportunities to craft unique customer experiences, optimize business practices, and stay ahead amidst the ever-evolving corporate landscape.

Chapter 9. Career Prospects in the AI Era

As we step further into the era of Artificial Intelligence (AI), an array of new job opportunities is materializing. Changes in the job landscape can often yield uncertainty or trepidation, yet they generate unprecedented professions and areas of expertise, widening the horizons for career prospects.

9.1. Jobs Directly Related to AI

The AI industry is constantly expanding, creating a multitude of job roles that directly interact with AI. Some of these job roles have traditionally existed, while others have arisen as a direct response to the burgeoning AI market.

1. **Data Scientists:** Even though this profession isn't new, it has become more essential than ever in the AI era. Their job is to extract valuable findings from raw data, a crucial task in the AI ecosystem, which relies heavily on large amounts of data.

2. **AI Engineers:** These professionals take the feasibility study and model building done by data scientists to the next level by coding and implementing AI models. They require deep an understanding of AI algorithms.

3. **Machine Learning Engineers:** These individuals design and build machine learning systems. They are usually proficient in several programming languages and have an understanding of both software engineering and data science.

4. **Robotics Scientists:** This role facilitates the creation of physical robots that interact with their environment. Robotics Scientists employ Artificial Intelligence to design and create machines capable of performing various tasks.

5. **AI Ethicists:** As AI technology continues to evolve and increasingly affects our daily lives, issues of ethics and morality in AI application grow. AI Ethicists, therefore, work to ensure that AI technologies are developed and used within ethical and legal parameters.

9.2. The Importance of Soft Skills in the AI Era

The evolving AI landscape isn't only creating opportunities for those with technical expertise. Companies are identifying the importance of integrating AI with human skills. This convergence has resulted in an influx of roles that prioritize soft skills. Some key soft skills include:

1. Communication: No AI can match the complexity of human communication. Effective communicators who can explain complex AI concepts to clients and colleagues are in high demand.

2. Emotional Intelligence: AI technologies lack the unique human capacity for emotional empathy, making this a crucial skill in the AI era. For instance, in healthcare and counseling, AI can be beneficial, but it requires the human touch to provide empathy and emotional support.

3. Critical Thinking and Innovation: AI is excellent at repetitive tasks but struggles with original thinking. Innovative minds who can think beyond the status quo will always be critical to advancing AI.

9.3. Educational Opportunities in AI

The advent of AI has spurred changes in the education system to prepare individuals for new job requisites. Universities and colleges now offer specialized courses and degrees focusing on AI, Machine Learning, and Data Science.

In addition, online platforms offer certificate courses for professionals looking to pivot into AI-related careers. Even young students are given a chance to explore AI, with some high schools introducing AI in their curriculum.

The AI era doesn't just create new jobs; it widens the scale of education and learning for everyone across different ages and job domains.

9.4. AI Impact across Various Industries

AI has permeated a host of industries, leading to a broad expansion of job prospects within each one.

1. **Healthcare**: From aiding diagnosis to enhancing patient care, AI is transforming healthcare. It is creating a demand for professionals who can marry medical knowledge with AI proficiency.

2. **Agriculture**: AI is being utilized for crop monitoring, forecast-based farming, and predictive analytics, opening a new domain for agricultural scientists and AI developers.

3. **Automotive**: With a future promising autonomous vehicles, a range of opportunities has emerged for AI engineers, robotics experts, and data analysts in the automotive industry.

4. **Retail/eCommerce**: AI is being used for personalized marketing, inventory management, and customer service. Consequently, roles for business analysts, data scientists, and AI developers are ramping up in this sector.

By interweaving itself across sectors, AI fosters a demand for both technical and non-technical roles, amplifying the career prospect pool.

As we navigate further into the AI era, the AI landscape continues to evolve, creating a dynamic and exciting playground of job opportunities and career paths. While the AI revolution has certainly obsoleted some jobs, it has also created many new ones. This continuous shift is not a cataclysm but part of an undergoing evolution - one where those who can stay adaptable, innovative, and keep learning will thrive. AI has warmly opened its doors to anyone who is willing to grow with it. The trick lies in staying agile, embracing change, and relentlessly innovating. In the AI era, the world truly is your oyster.

Chapter 10. AI and the Individual: Everyday Implications and Interactions

Artificial intelligence (AI) has gradually permeated everyday life, reshaping our routines, tasks and outlooks. From our smartphones to our workplaces and even our homes, AI applications are becoming so integrated into our daily routines that we sometimes overlook their presence.

10.1. Understanding AI in Everyday Life

The modern age has witnessed AI evolve from science fiction fantasies to reality. Today, AI isn't merely about robots replacing human jobs; it's far more subtle and pervasive than that. Everyday AI involves algorithms analyzing our online behavior, virtual assistants organizing our schedules, predictive technologies shaping our shopping experiences, and much more.

Chatbots and digital assistants like Siri and Alexa are entrenched in our lives, aiding in task management from sending emails to scheduling appointments. These AI-powered assistants understandably generate an impression of personalized interaction, which contributes to their wide acceptance. Real-time traffic updates on our GPS systems rely on AI algorithms to display the most efficient routes, optimizing time and minimizing fuel consumption.

When we shop online, AI comes into play yet again. Algorithms analyze our browsing patterns and preferences, providing personalized product suggestions. These predictive algorithms not only enhance customer experience but also streamline business operations.

Yet, the implications of AI go beyond pure convenience. It has made it possible to create predictive models for health care, personalized learning plans in education, and even advanced security systems. Let's delve deeper into these sectors.

10.2. AI and Personal Health Management

AI's presence in healthcare is one of the most profound areas where it impacts individuals' lives. From wearables that monitor our health metrics to AI-powered digital platforms that provide personalized fitness advice or medical information, AI is revolutionizing the way we approach wellness and disease treatment.

Smartwatches and fitness trackers monitor vital signs like heart rate, blood pressure, and sleep patterns. They provide us with relevant data that not only allows us to track our fitness progress but also predict potential health issues. AI-powered bots such as Babylon Health offer medical consultation based on symptoms entered by the user. The AI-driven software Ada goes a step further and provides a comprehensive health assessment.

AI applications also extend to mental health, with apps like Woebot providing cognitive behavioral therapy exercises. The potential of AI in healthcare is immense and could change personal health management radically, leading to enhanced wellness and longevity.

10.3. AI in Education

AI is similarly transforming education, bringing personalized learning to everyone. AI algorithms assess strengths, weaknesses, and learning styles of individual students and devise tailor-made learning paths.

AI has enabled the creation of adaptive learning platforms like DreamBox Learning, which assembles learning resources based on each student's knowledge level and learning speed. This personalized support compensates for the lack of individual attention in traditional classroom settings, making education more equitable and potentially more effective.

Besides, AI applications such as automated grading of multiple-choice tests or AI-driven plagiarism detection improve the efficiency of administrative tasks, allowing educators to focus more on teaching.

10.4. AI and Personal Security

The advent of AI also brings with it a significant improvement in personal security. Facial recognition technology, powered by AI, is now commonplace, used in everything from unlocking smartphones to identity verification. These technologies owe their origin to machine learning, an important subset of AI, where computer systems are trained to learn from data and make accurate predictions.

AI also enables more secure financial transactions. Banks and financial institutions use AI algorithms to detect fraudulent transactions. AI can analyze countless data points in a fraction of a second, something human beings can't do, making this technology invaluable.

However, with these advancements comes the question of privacy – a topic we will explore later.

10.5. The Possibilities and Challenges Ahead

Despite the impressive applications we've reviewed, we've only just tipped the iceberg regarding AI's potential impact. AI could dramatically change transportation with autonomous vehicles, revolutionize the way we consume media with personalized content recommendations, and enhance our connectivity with smart homes.

AI also offers the potential for significant advancements in energy efficiency. AI algorithms can optimize power usage in appliances, potentially saving us millions of dollars while reducing carbon footprints.

However, these advancements aren't without challenges. Privacy is a key concern that arises alongside AI's triumphs. The data that fuels AI systems is often personal, and its handling can lead to potential breaches of privacy. Also, job displacement due to AI automation is a valid concern that societies face.

It's crucial to navigate these challenges conscientiously as we move forward. Regulations and guidelines must be put in place to protect privacy and ensure ethical conduct. Jobs will undoubtedly change as AI automation increases, but new roles will also emerge, roles that we can't envision today as we stand on the precipice of this exciting era.

Indeed, the future of AI is bright and holds countless opportunities. As we continue to develop and refine this technology, and as we learn how to ethically gather and use data, we'll see unprecedented advancements in everyday life. AI's story is still being written, and we're all part of it. We are in a unique position to shape this narrative, ensuring it's one that benefits us all.

Chapter 11. Conclusion: Preparing for the AI-Driven World

As we conclude this in-depth exploration of the dynamic environment of artificial intelligence, it becomes necessary to equip ourselves with constructive strategies and paths that will enable us to navigate and even leverage the AI-driven world we will increasingly find ourselves in. Keep in mind, the future is not a standalone concept; it's shaped by the choices and decisions we make in the present.

11.1. Embrace the Changing Reality

The adoption and normalization of AI are not distant, futuristic events; they are already steadily transforming our society. From the AI driven recommendations on e-commerce websites to autonomous vehicles, AI is vigorously manifesting itself across both industry and daily life, resulting in a wave of disruptive innovation. To stay relevant and seize the opportunities that lie ahead, both organizations and individuals must adapt and become fluent in the language of AI.

Understanding the basics of AI, acknowledging the importance of data for machine learning and appreciating ethical aspects of AI are initial but essential steps in this process. Whether through formal education, online learning platforms or on-the-job training, gaining foundational knowledge of AI will enable us to make informed decisions.

11.2. Fostering Digital Literacy and AI Education

With the increasing use of AI in diverse fields, we can no longer afford to restrict AI understanding to specialised fields or tech companies. Countries and organizations should strive for a widespread digital literacy that includes AI. Education systems need to introduce students to AI concepts from an early age, cultivating curiosity, promoting understanding and instilling responsible usage of AI technologies.

There is a need for digital skills – ranging from basic digital literacy, digital skills related to jobs, and advanced digital skills such as coding, data analysis and AI literacy. Through this comprehensive skill-set, the next generation will be able to navigate the increasingly digital world more successfully and even influence the AI landscape of the future.

11.3. Nurturing AI Innovation

Continuous innovation in AI will lead to new solutions and products, reimagining the way we interact with the world while driving economic growth. Encouraging entrepreneurs, startups, research institutions, and corporates to build on AI platforms and embark on AI-focused projects will stimulate this innovative culture.

Economic policies that foster AI development, paired with investment in relevant public infrastructure, can lead to an ecosystem allowing AI to flourish. We must understand, however, that the best environment for AI innovation is one that encourages risk-taking while also supporting those who may initially fail, enabling them to grow and learn from their experiences.

11.4. Ethics and Regulations in an AI World

As AI becomes more integrated into society, ethical considerations and regulations concerning data privacy, fairness, accountability, and transparency will become ever more critical. Striking a delicate balance between regulation and innovation is essential so as not to choke AI innovation while also protecting society's most vulnerable.

Public and private sector leaders need to work collectively to build an agile, adaptive regulatory environment that monitors and guides AI development and usage. Meanwhile, AI ethics education must be mandated in all AI-related educational programs to ensure the developers of tomorrow build AI systems that respect fundamental human rights and values.

11.5. Mitigation of AI Induced Displacement

AI has the potential to automate certain jobs, causing displacement. While efforts must be concentrated on upskilling and reskilling the individuals potentially affected, we also need to explore the creation of AI-driven jobs and new types of work. Identifying the areas that will likely be automated will guide effective workforce planning and proactive human resources strategies that ensure the wellbeing of the workforce.

11.6. Collaborative AI Advancements

The advancement of AI should not be a race but a collaborative effort. Different nations, companies, and individuals have unique perspectives and insights, which, when shared, can lead to a more comprehensive and beneficial AI development. Global cooperation on AI development and regulation will prevent the creation of uneven power structures and ensure that benefits are not limited to select regions or factions of society.

In conclusion, as we journey into an increasingly AI-driven world, we need to equip ourselves with the relevant knowledge, nurture innovation and foster a globally inclusive AI community. We must advocate for robust ethical practices while addressing AI-induced displacement through proactive planning. Above all, we need to recognize that the future of AI is a collective journey that requires our mutual collaboration and shared vision. As we shape and influence this world of AI, it is paramount that we do so with an unwavering commitment to promoting shared prosperity and respect for human dignity. The AI future is indeed a remarkable adventure, and with the right preparation, we can turn those striking revelations and exciting opportunities into meaningful, positive outcomes.